HOW TO ANALYZE PEOPLE IN 5 MINUTES

The Ultimate Guide to Read People in 5 Minutes or Less. Through Psychological Techniques, Body Language Analysis, Reading Personality Types and Predicting Human Behaviour

By Nick Reynolds

© Copyrighte 2018 – All Rights Reserved.

All rights reserved. No part of this publication may be reproduced, distributed, or transmitted in any form by any means, including photocopying, recording, or other electronic methods without the prior written permission of the author, except in the case of brief quotations embodied in reviews and certain other noncommercial uses permitted by copyright law.

Disclaimer Notice

Although the author and publisher have made every effort to ensure that the information in this book was correct at press time, the author and publisher do not assume and hereby disclaim any liability to any party for any loss, damage, or disruption caused by errors or omissions, whether such errors or omissions result from negligence, accident, or any other cause.

Introduction

Chapter 1
Words are powerful "Understanding Word Clues"

Chapter 2
Setting Standards and Baseline

Chapter 3
Analysing Someone's Personality

Chapter 4
Analysing Personality Traits

Chapter 5
Personality Traits that Predict Ideas

Chapter 6
Errors in understanding motives

Chapter 7
Secrets of Reading Body Language

Chapter 8
Quick signs that someone is lying

Chapter 9
Judging a book by the cover

Chapter 10
Emotional Bullying, How to spot it!

Chapter 11
Analyzing People by Birth Order Behaviour

Chapter 12
How to detect romantic attraction
Putting it all together

Introduction

We all would like to analyse and read people on spot, it would save us a lot of trouble, a lot of disappointment. Understanding your co-workers, your friends and strangers once you meet them. Knowing who is who, how to deal with them, manage the relationship and knowing who to trust and who to share anything with.

We all would like that superpower to read people's mind, I would too, it would make life easier in understanding my wife and what in the world would they really like to eat. Unfortunately this is not possible ever, but what is possible is for you to understand people behaviours and their personality, by what they say, their actions and a few other queues.

There is nothing more interesting in understanding why people do what they do, people are very interesting and complex. I often ask why didn't God make more person like me, often you interact with the person and it makes you really wonder. People are peculiar, sometimes weird, for instance, my neighbour she cuts her own grass by herself and she is about 60 years old, why? I have a friend that cooks his food every day and brings to work, even though he can afford to buy from any restaurant, why? Why would your co-worker turn up the A/C then when you ask them to turn it down, they refuse, knowing all of the rest of their co-workers are freezing. Thinking only for themselves, yet they expect others to be considerate about them.

There many things that just leave us confounded at the times, I mean just confused as to why would people behave the way they do right? When we can analyse people, we can predict and understand their actions. This will make dealing with people so much easier and also a bit fun. The challenge we have in life is not dealing with people, but in understanding how to deal with them. We have to interact with people, but the quality of interaction that we have with them can be determined by how we much understand them.

When we can analyse people more effectively, we can figure out who are the best person to keep as friends, who are the best person to go into business with. Who are the best person to be in relationship with and who are the ones to

just cut out of our life immediately.

Scientist has discovered many ways in which you can analyse people and their personality. You can use this information to predict their actions and how they will react to certain situations.

You can use strategies such as word clues, listening to the words that people use, and the terms can say a lot about who they are. Not only who they are but what are their thoughts, how they think, process information and interpret information. When you can understand word clues you can predict their behavioural characteristics, not entirely but it can give us some insight as to how to they will behave.

We can also have more insight about a person by using a baseline, a baseline is something that usually occurs based on a certain stimulus. Or a reference point, for instance, if you have a person that behaves in a particular way towards the conversation about animals it is easy to develop a baseline from that interaction. If an individual becomes very irritable or sad when having conversations about a specific topic then from that you can developed your baseline when interacting with that person. If you see where a person becomes very unsettling about the conversation about relationships and intimacy, from that point you can develop a baseline and an idea about the person. You may have met a person that are less open to going in a relationship, less open to courting. From these baselines we can get an idea that this person may have been broken by some circumstances as a result they have loss trust in everyone.

These insights can be gain when we have baselines, baselines are fundamental to understanding any behaviour about any individual that we are interacting with. Let's look deeper into how we can break down in depth ways how we can read people.

Chapter 1
Words are powerful "Understanding Word Clues"

Words are powerful, they represent thought and intent and words reveal a person's behavioural characteristics. Whether they are written or spoken they convey who we may be in the same way. Some may say that people tend to hide their true intentions by putting up a front, or misleading person through words. That can be so if the act is deliberate, however through other means and mechanism, you can still analyse who a person really is. Insights can be gain from the words people use, as they reveal their thought process.

A hypothesis can be developed from the word clues you receive from using additional information gain from further interactions.

Word Clues allows you to read anyone without their knowledge or forewarning, which can effectively predict their behavioural characteristics.

Let's dive into some examples as to how we can analyse people by their words.

"Hey, guys I won another prize!"

What if you were to hear one of your friends or someone made this statement, obviously by you hearing them simply saying "another" shows that they are trying to indicate that they have won a prize before. This word clue shows that this person may like the recognition that comes with being awarded. They may need adulation that would reinforce their self-image. These type of persons love the admiration of people and they may be a vulnerability, persons may take advantage of this at times. But from this statement, as you can see, the person love admiration, and adulation, you can use this to your advantage or otherwise.

"Every day I worked really hard to achieve this"

What would be the word clue here?.......................... That's correct "Really hard" this shows that this person value, their achievement, value goals, that are challenging to achieve. It could be that they extended extra effort on this specific goal. The word clue "really hard" may also suggest that the person defer many gratification, and made sacrifices. If you are an employer, a job applicant with these characteristics would make a great employee because of the fact that they are determined to be successful and complete their tasks.

"I patiently sat through the tutorial"

Patiently here, as a word clue, could suggest any of several things. It could be that the person had to sit through a boring lecture, it could be that the person had something else that is pertinent to do. Perhaps they wanted to leave to use the restroom, or to grab something to eat. A person who normally waits patiently is often a person that adheres to social norms and etiquette. This shows this person may have rigid social boundaries, these type of person follow rules and respect authority.

"I decided to buy that"

The word clue here "decided" provide an idea that the person may have weighed other options before settling on an idea. The person may have struggled and deliberated a lot before making the purchasing decision. This person thinks through before making a choice or a decision. An impulsive person may say "I just bought that". The word clue "just" suggests that the person may have made the purchase without much thought.

The fact that the person said that they, "decided" means the person may be introverted. Introverts normally think before they act, they weigh all their options before making decisions. While on the other hand, extravert tends to be more impulsive and spontaneous.

Extraverts often get their energy from interacting with people and outside environment. On the other hand, introverts like to spend less time socializing and more time by themselves. The like to get their energy from within and like to find time to recharge. Introverts carefully weigh their options before they make any decisions. Introverts take information and analyse it, mull it over and then make their decisions.

Extrovert, however, are comfortable with making impulsive decisions. It's not often that you see a 100% introvert or 100% extrovert. They often display different personality based on the situation, but one trait is often dominant. Extravert can display introverted behaviour and an introvert can display extraverted behaviour.

"I am sure what I did was the right thing"

This phrase, with the word clue "right thing", may demonstrate that the person struggled over the thought of moral, ethical and legal. They may struggle with the thought to make a just decision over a dilemma. This trait shows that the person has enough character to make just decisions. This person may be a person that you can trust, that you can go into business with, because of the fact that you know they will make morally right decisions.

Listen to the words person Re-Use

"That machine is really powerful" "That colour is very strong" "I don't have the strength to go to the gym". All these statements are related to some extent, if you pay attention, you will see that the person has mentioned the word power or synonym of power in each phrase.

The vocabulary inside the mind is not random, the words we say are constraint by our core needs, desires, ambition, and concerns. The person who said the words above is obviously concerned with power and weakness. The person may want to be strong, powerful or even wealthy.

Listen to how they tell stories

If you hear them telling stories to you or their friends, using the words huge, or powerful. For example a story like this "Last night I was walking down the road and saw this gigantic man standing in front of me, it seems as if he was trying to do something to me". From that sentence, you can tell that the person may have a concern about power, the person may have a desire to be strong, it could be emotional as well. The fact that the person may have left out all the details and focus on the fact that the person was "gigantic", shows that that was what concerned him the most.

Listen to their jokes

When you listen to people jokes and their comments, you can pick up on their hidden messages. You may here person make jokes saying, they would like to drive a Porsche or they would love to be in a mansion. From these comments, it is safe to say that this person would really like to become rich, their background, and their family may have had some financial issues, or they may grow up poor. These are just a few examples as to how to read people from their words, statements, and comments. When you meet a person, and you hear them making a few statements and using similar words, from that you can develop an idea about the person. Only judging a person from a one-off statement may not do, however, when you engage in conversation with the person, you will get a general sense about who they are.

Chapter 2
Setting Standards and Baseline

Baselines are the standards by which, you will use to measure other event and changes. The Baselines will be used as a reference point that you will use to make all other judgments. For example, if you are doing a research in the rate of increase of sugar consumptions, you would first need to collect information for your data. The data that you now have, that you will use as a measuring stick, will be your baselines.

Some people may call them to reference point, anchors, adaptation levels, but it's the same baselines. We all have them at some point in our lives. Most of our baselines are informal, but these baselines are also set up to protect us often time. For example, a person may look a little threatening as if they may grab your purse, or try to steal from you. A lot of times, we may be wrong, but other times we are right and we have to take steps to protect ourselves. Baselines help to shape our lives, push us forward and protect us from some actions we would otherwise take.

Some may call it instincts, however, instincts are something we develop over years of experience and observation. It's just the same as the example I have given you before. For example, you have times, by just interacting with a person once, you will walk away knowing that the person was a nice person, or a bum just from one interaction. That is because you have a baseline for who a nice person is, and who a bum is. Not all the time, but often our interpretation is correct. Our ability to read a person body language in a given situation will allow us to develop a baseline about that person in a specific situations, for example, just by how a person reacts in a stressful situation, we can then make an assessment of how a person may deal with stress. Just the same as cracking personal jokes, all these baselines can be developed from these interactions.

Chapter 3
Analysing Someone's Personality

There are a lot of ways you can understand people, or understand their personality, some are through reading body language, facial expression or even how they talk and walk. There is some truth to this, however analysing a person's personality require a bit deeper knowledge than what is there on the surface. While I was browsing the internet, I came across a story on a website of how someone figured out a person just by the way they behaved in the gym. Imaging that, your ability to analyse people so well, that even when you are not directly interacting with them, you can read and understand their personality.

The writer said that he noticed that the person came to the gym, every day at the same exact time, he noticed the gentleman was self-motivated, not a procrastinator and very strong-willed. He also noticed that the person was an attention seeker, he used to wear revealing clothes that hugged his muscles and do heavy weighted exercise, he was a bit of a show-off. After the writer collects more information about the guy, he found out that the guy was the only child. These type of persons like to remain at the centre of attention, a bit spoiled.

It is important to note how the birth order affects personality, birth order is the order in which the child was born this will often determine their personality trait.

The writer also noticed that another guy often wear black, all the time, even when he would change his clothes they would still be black, his phone and the case were black as well. From that, he could tell that the person was trying to appear tuff to everyone. After the writer collected more information he discovered that the person was actually bullied in school. Because he grew up weak he tried to do everything to cover up his past.

What we also have to identify is that when we try to connect the dots when analysing a person, they all must add up and point back to the same direction. Meaning, what a person may be doing in another facet of their lives will contribute to the same thing of being powerful and strong.

Knowing someone personality does not take a long time, once you spend enough time around them. You can arrive at your conclusions quite easy. All you have to do is pay more attention to the person's behaviour and the tiny actions that people mostly overlook. Subtle things such as the fact that a person may love to go on adventures, rock climbing, bungee jumping, skydiving. This proves that the person may be comfortable with taking a risk, even if it was not physical. These people you can identify would be fun, spontaneous and may enter into businesses and entrepreneurship. This type of person can quit their job at any time, without no second thought. In the other end, there may be some person, who like to sit at the same place every day, even when he goes out, They would often go to the same places over and over again. This type of person is said to fear change and his apprehensive to doing something new.

While a person who may take their time and look both ways multiple time before crossing the street, maybe a bit hesitant in taking a risk or slow to making decisions. These type of persons often take calculated risks or take time evaluating all the pros and cons before they make a decision. You will be able to better navigate through life if you really understand people personality trait.

Chapter 4
Analysing Personality Traits

There are several personality types, when you understand them, you can make better assessments of people. As it relates to understanding their personality, who they are and how to deal with them. The four basic personality type identified by Car G. Jung has allowed classifying an individual in categories that generally define and predict their behaviour.

The Basic functions are Sensing (S), Intuiting (N), Thinking (T) and Functioning (F) in either the external (extrovert) or internal (introverting). Carl G. Jung used 8 cognitive processes and express them in capital letters

(S, N, T or F) plus a lower letter "e" (extrovert) or "i" (introverting) to indicate orientation. So (Se) where (S) is sensing and (e) is extrovert, so (Se) means extrovert sensing. Sensing relates to the relationship with the outside world.

Author Isabel Briggs Myers and her mother, Katharine Cook Briggs, created the Myers-Briggs Type Indicator (MBTI). A test that mad Jung's theories more user-friendly and understandable.

People differ in many ways, in their perceptions and how they arrive at conclusions. With that we expect that to also differ in their interest, reactions, what motivate them and their skills. This theory is according to Myers-Briggs Foundation.

The Myers-Briggs Type Indicator is not the "be all and end all", but it will definitely turn on some light bulbs and help you understand and tell personality type better. This information provides you with eyes to the kind of understand what's going on with a person and how you can predict their behaviour. It allows you to really approach things differently in terms of dealing with people and approaching them, differently.

The Myers-Briggs Approach

Here is what the Myers-Briggs Approach really looks like:

If you prefer focusing on the outer world, this is called extraversion or "E", whilst if you prefer focusing on your inner world you are considering Introversion or "I".

If you prefer to focus on the basic information you are more (Sensing or "S") or do you prefer to in interpret things and find meaning this is considered as (Intuiting or 'N")

Are you a person that first look at logic and consistency (Thinking or "T"), or you may first examine the circumstances and people when making decisions, this is considered (Feeling or F)

When you are dealing with the external world, if you rather prefer things decided (Judging "J") or if you prefer to be more open to options and information you are (Perceiving or "P").

There is 16 personality type of the MBTI, they are based on a combination of preferences in each category, expressed as a four-letter code, and however, the MBTI does not measure traits, ability or character. A psychologist has noted that the goal of understanding personality types is to have a better understanding and appreciate differences among people.

Let's examine some different personality type from the Meyers Briggs Personality type.

INTJ PERSONALITY ("The Architect")

These personality type INTJ represent less than 2% of the population, the INTJ is rear and it is hard to find people liked minded. They have a different level of intellectualism and they have a chess like manoeuvring. These personality types are imaginative, yet they are decisive, they are also very ambitious and private at the same time. They are very curious but they know how not to squander their energy.

The INTP personality type is fairly rare, making up only 3% of the population, which is definitely a good thing for them, as there's nothing they'd be more unhappy about than being "common". INTPs pride themselves on their inventiveness and creativity, their unique perspective and vigorous intellect. Usually known as the philosopher, the architect, or the dreamy professor, INTPs have been responsible for many scientific discoveries throughout history.

INTP Personality ("The Logician")

INTP personality is a very brilliant theorist, and are considered as the most logical personality types of all. These type of people love patterns and spotting mistakes and discrepancies. These people often like to share ideas and develop them, using other people as a sounding board for ideas. These set of people, they are extremely enthusiastic and keen to spotting problems, getting down to the nitty-gritty details.

The INTP personality type, aren't interested in practical day to day activities, monotonous jobs. The seek environments where they can be creative, where they can express their genius with no limit on time and energy. INTP thought process is active from when they are awake. They are always thinking of conducting debates in their heads, some of them are quite shy and relaxed if they are around familiar faces. Some of their conversations may be a bit incoherent as they try to make sense out of their logical conclusions.

ENTJ Personality "The Commander"

These type of people are natural born leaders, they have the gift of charisma and confidence. They know how to rally people together to achieve a common goal. Sometimes they can come off as ruthless, using their drive and determination to achieve whatever that they want. These type of people make up less than 3% of the population, ENTJ often becomes very successful at whatever they do. The ENTJ love challenges, they have a deep belief in themselves that they will overcome any challenges they have with enough resources and time. These type of people are very strategic, they are committed to their goal, super focus while they execute their plans. They have an extroverted nature, they often push everyone along to support their objective, they are often vicious and relentless.

Some ENTJs can display some level of arrogance and behave in a condescending manner, which will not go over well with everyone. Some people may often shy away from ENTJ's because of their personality.

INFJ PERSONALITY "THE ADVOCATE"

These personality types are very rare, there is less than 1% of the population. However, because of their personality, they tend to leave a mark on the world. INFJ are diplomats, they have a sense of idealism and hold strong morals. They are not mere dreamers, they take actions necessary to achieve their goals. INFJs are very helpful, they often mentor people in finding their purpose in life, and they are also active in charitable work. They have very strong opinions and will go to the ends of the earth to defend their ideas. They are strong-willed humanitarians. They are selfless in their objectives, with strong convictions, these type of people often change the world around them.

INFJs they can connect with people easy, they are warm and speak in human terms, rather than try to impress or show off their eloquence. They are a bit of extraverted as well, but they really like time for themselves, where they can sit back and just recharge by using up their imagination.

ENFP PERSONALITY "THE CAMPAIGNER"

The ENFP is often the life of the party anywhere they go. These people, are a true free spirit, they have a sense of adventure, and they are more interested in the pleasure of sharing a moment with other people in a social environment. They are also charming, independent and energetic, a small percentage of the population actually comprises of the

ENFPs. But anywhere they go, their personality can definitely be felt in the space. They believe they can change the world with their ideas, these type of people are 'social people pleasers'. They are Diplomats that are shaped by their intuitively. They are able to read between the lines and tend to see life as a big, complex puzzle interconnected. They see the world through their eyes of emotion, compassion, and mysticism, always looking for deeper meaning to what naturally occurs. ENFPs will bring high energy to creative ideas, an energy that will often have them in the spotlight. They like to be revered by their peer as a guru type leader. Their self-esteem is really high and is somewhat dependent on them coming up with solutions to problems. They need to have the freedom to be innovative and creative, they can quickly get impatient and bored if they are doing the same monotonous task.

This is the most popular type of personality, making up approximately 13% of the population. ISTJs they respect traditions, help to uphold them, they are vital to family and organizations. These type of personality like taking all the responsibility for their actions and have high regard for their work. They are more focused on their goals and see through each task until completion with patience and accuracy.

They are also a bit analytical, they make little or never at all make assumptions, and they like to make many checks before they take actions. They are the type of no-nonsense personality, once they have made a decision, they will communicate the same to other expecting them to buy into their decision and follow through immediately.

ISTJs are decisive and have very little patience if the direction that they have chosen, which may be later found to have many impracticalities. If their task becomes a series of debates and deliberation, this will see them become a little agitated as the deadlines become close.

The ISTJs are a form people that, when they decide to do something, they do it. They hold them self to a higher standard in terms of their obligations, and they are baffled when they see people that don't do the same. They frown upon laziness and dishonesty, they disassociate themselves from such people. They like to work alone, where they can be accountable to themselves and don't have to worry about reliability.

ISFJ PERSONALITY (THE DEFENDER")

The ISFJ personality is a bit unique and peculiar, even down to their qualities which defers from their traits. Even though they possess the Feeling "F" traits, they also have exceptional analytical abilities; they are Introverted (I), yet they have people skills and are good with social relations. They are the Judging (J) type, they are often receptive to change.

ISFJs are very altruistic, they are very kind and they believe in working with people that are generous and enthusiastic. And it's such a good thing that they make up nearly 13% of the population. You often find them in careers such as medicine, academics, and charitable work. Perfectionism is a part of their core, though often they will procrastinate, they will execute the work on time. Doing anything that is necessary to delight others and exceed people expectations.

The ISFJs like the fact that their work is noticed, that they are acknowledged for that they are doing. They are very humble and dedicated, this can be to their disadvantage as other personalities that are greedy may take credit for what the ISFJs have done. They are introverted yet naturally social, they are excellent at remembering details about personal lives.

ISTP "The Virtuoso"

These personality love to explore with their assets, eyes, and hands. They find the world around them interesting and that piques spiritual curiosity. They are superfluous, they are Makers of things, find joy in moving from project to projects. ISTPs are similar to engineers and mechanics, they enjoy digging down, and getting their hands dirty. They love to pull things down, and fix them up back together kind of what we were like when we were a kid.

The ISTPs they like to conceptualize their ideas, by troubleshooting, hands-on simple trial, and error.

They love when people take notice in their work with intrigue and amazement. They enjoy as well helping out and offering advice to others they think need it. They make-up 5% of the populations, very enigmatic people, yet they still respect and love their private time. They also find it really difficult to focus on formal studies. These personalities are not that easy to predict, even by people that close to you. They are heavy explorers, they are interested in making big shifts and changing the pattern of which they were going.

The biggest disadvantage with ISTPs is that they act on impulse, they are the first say something insensitive and get involve in other person projects. ENTJs are a distance from emotions especially in public or even in a professional environment. They are really authoritative and have an image in their minds, that they are larger than life. They believe that it's by their efforts that they have achieved all that they have.

ISFP "The Adventurer"

ISFPs are real artists, they use aesthetics and design to push against limits and social convention. They love to go against the grain and upset traditional expectation.

Inside the mind of an ISFPs, they live in a colourful world, sensual and connected with many ideas. They find a thrill in reinterpreting what they perceive in the world, they like to reinvent and experiment with themselves and create a new perspective. Once you get to know ISFPs they will seem unpredictable and have a high level of spontaneity.

What may be surprising is that ISFPs are actually introverts, they like to step out of the spotlight and recharge, even though they sit and look idle, they are active. Active doing introspection, they tend to think about who they are and what meaning they have to live. The ISFP they often do risky things or partake in risky behaviour. They will participate in things like gambling and extreme sports. They may not take kindly to people giving them comments against their risky ideas. They may tend to lose their temperament and it may be even chaotic.

ISFPs know how to empathize with other people and they value peace more than anything else. They are the type that lives in the moment and may find it hard to walk away from a heated argument. Which play to their advantage somewhat, as when they walked away from something they will just leave it in the past and move on.

THE BIG FIVE MODEL

One of the ways to analyse a person personality trait is through the BIG FIVE model. I am sure most of us were exposed to this during our school years. The BIG FIVE model is alternately called CANOE or OCEAN; which stands for Openness, Conscientiousness, Extroversion, agreeableness, and neuroticism.

The first word in the acronym openness, which explains itself. This is for persons who are more open, they explore, try to do new things, high curiosity. These type of people have a broad range of interests, they are very curious about their surroundings and other people. They are enthusiastic to wake up and enjoy new things and experiences.

People with openness traits tend to be more imaginative and have more insight, these type of people have a broad range of interest and curiosity. They are curious about what is happening in the world, and its environment. They are eager to learn things and enjoy more experiences, they have a high need for adventure and creativity.

The E of OCEAN is for Extroversion, the extroverts like stimulation, they like to engage in things that bring, things that are engaging and exciting. They are more energetic and are willing to take over in situations, especially social ones. They are willing to take lead in many situations especially social events and activities. Extraverts feel as well that they can handle anything in life that is thrown at them, it is their proactive approach that gives them that belief and confidence.

A is meant for agreeableness, these personality traits include characteristics such as trust, altruism, affection, love, and kindness. People whose personality trait is Agreeable they are more cooperative and inclusive. They have a high deal of interest in other people, they care about others very much, and these people are very affectionate. Persons with these personality look for social harmony, try to avoid conflict at all times. Agreeable people also have an optimistic view about humanity, they believe that people are generally honest and trustworthy.

Agreeableness people are generally very popular, because of their concern and empathy with people. The disadvantage is when they have to make tough, objective decisions. Hence often they do not do very well in roles that require huge responsibility.

The last letter in OCEAN is obviously the N, which stands for Neurotic. These personalities are inclined to experience negative emotions, anxiety, depression or vulnerability, sometimes referred to as emotional instability. Persons with Neurotic personality trait, tend to worry more, they are very insecure and self-conscious. Neurotics are a very fragile personality, they are often at risk for depression and may at times be suicidal. Studies have shown that persons with neurotic personality trait account for 32% of the suicide rate. These type of person are very vulnerable, thus we have to be very careful of how we manage our relationship with these personality type.

Different Personality Types

By having a knowledge of the different personality type, and knowing how to identify them, this will allow you to understand who you are dealing with and how you should manage that relationship. There are 8 cognitive functions, and these all depends on what letters make up your personality. Let's have a look into the different cognitive functions, based personality type.

Extroverted Sensing (Se)

These type of person do better in stimulating environment with quality sensory input. They like their surroundings to be open, lots of windows, exciting views and décor. When you are around them, they are very encouraging and help to drive productivity and resourcefulness. They prefer to live in the present moment rather than dwell on past events. They are very Intune with their surroundings and environment, extremely self-aware.

Introverted Sensing (Si)

Introverted Sensing people much rather little distraction, and time to reflect on past experiences in order to make future decisions. When you are working with these personality types be careful of how you provide them with feedback, be aware of nonverbal clues from them. Which will tell you how they received your information. These individuals rely mostly on stored memories and experiences, anything that they experience in real time, they immediately make reference back to something they experience in the past. In essence, Introverted Sensing personality has a high comparative attribute to them, their brains are wired to make comparisons. They have the ability to call back on past experiences to make decisions, they can remember things in great details as well.

Introverted Intuiting (Ni)

Introverted Intuitive people sometimes need to spend time away from everything in solitude in order to pull on their resources from within. These people when they get their information from the external environment, they often don't deliberately or try to make sense of it. It is usually after the information has been processed in the subconscious mind that they arrive at whichever conclusion. They focus inwardly for most of their decision making as opposed to the extrovert that focus externally. (Ni) are adept at analysing events and experience from the past, they use them for clues, dominant (Ni's) are more focused on the larger vision, they are less detail oriented.

Extroverted Intuiting (Ne)

The Ne's are people that gather input from other people when they work, they don't mind working in distractions such as radio, televisions etc. Intuition is the perceiving functions, and Extravert has to do with how they behave, Ne involves obtaining information from outside sources. The Ne's prefer to adapt and to blend in with their environment rather than control them.

This allows Perceiving types to readily adapt to and blend with their circumstances rather than trying to change or control them.

While Se involves apprehension of information through one or more of the primary senses, Extraverted Intuition does go beyond or looks behind sense data. This allows NPs to discern otherwise hidden patterns, possibilities, and potentials. Extraverted Intuition scans for relationships or patterns within a pool of ideas, facts, or experiences. In conjunction with either Ti or Fi, it helps NPs formulate and modify ideas. NPs commonly employ their Ne in activities such as reading, conversation, and engagement with nature or the arts.

Extroverted Intuiting (Ne)

Extroverted Intuiting (Ne) people prefer diverse inputs for brainstorming, so when working with them, allow sensory distractions with television, radio, and friends. Allow their goals to coalesce from various inputs, mental processes, and side-tracks rather than pushing a linear process. Focus on meanings and relationships between ideas, making sure that analogies work well. And use some humour, wordplay and similar cognitive games.

Extroverted Thinking (Te)

Extroverted Thinking (Te) people use their brains in an energy-efficient way, relying chiefly on seeing measurable elements, hearing words and making decisions. They prefer to use and respond to facts and figures, and favour the use of visual/spatial formats like charts, diagrams, and grids. Do not mistake confidence and speed for competence.

Introverted Thinking (Ti)

Introverted Thinking (Ti) people tend to rely on sophisticated, complex reasoning, using multiple reasoning methods, including deducing, categorizing, weighing odds, etc. Their thought processes are not directly linked to sensory inputs, so decision-making tends to be deep and detached. Allow time for clarification, as this person makes and corrects mistakes, striving for high accuracy before implementation. Provide techniques to help deal with excessive social and emotional data, which may overwhelm them.

Extroverted Feeling (Fe)

Extroverted Feeling (Fe) people pay attention to your words and how you may evaluate them while showing very little outward physical signs of doing so. The ethics of people's choices and failings are highly salient to them, so allow room to discuss considerations of justice and injustice. Use and respond to value-laden language, focusing on word choice more than their tone of voice, which may remain steady even while upset.

Introverted Feeling (Fi)

Introverted Feeling (Fi) people listen intently, especially for a tone of voice, motivations, words that link to your values, and what's left unsaid. Speak thoughtfully, take your time and don't rush, because after listening, this person may seem surprisingly definitive about decisions. Speak to his or her values, especially positively felt ones, while staying true to yourself (no phony effect).

Chapter 5
Personality Traits that Predict Ideas

Personality is grouped in terms of traits, which are relatively consistent which influence our behaviour in a lot of situations. Traits such as conscientiousness, friendliness, helpfulness, honesty are helpful in as in they help us to understand the consistencies in behaviours. Here are some personality trait that predicts behaviours:

Authoritarianism

when you have encounters with people with authoritarianism personality traits, we will notice that they are conventional, superstitious, and have mental toughness. We have all had experience with someone with this personality trait, it's not often well received.

Individualism/Collectivism Trait

A person with the individualism trait, has the tendency to focus on themselves, their own goals. Whilst collectivism it's almost the opposite, this trait is where person focus on the relations that they have with others. Individualist would much rather do things that allow them to stand out from others. Collectivist will more focus on behaviours that highlight their similarity with others. Here is a list of individualist belief:

- They believe people must take care of themselves
- Self-Orientated
- They must make decisions based on individual needs

- Each person have a right to privacy
- They emphasize on and individual initiative and achievement

Here is a list of collectivism belief

- They expect loyalty in groups (family, organization, friendships)
- They make decisions based on what is best for the group
- Their identity is based on social systems
- They have a high dependence on belonging

Internal/External Locus of control

People with an internal locus of control are far more likely to believe that life happens based on their own efforts and characteristics. They are generally happier and less depressed than their opposite counterparts that are External Locos of control.

The opposite which is External Locos of Control, where a person believes that they have little control over what happens to them. They believe what happens is by design and is outside of their ability to influence the situation.

Need for achievement Nach

Individuals with a high need for achievement often have strong desires to make accomplishments by mastering skills. These individuals sometimes select an easy task that is less challenging so that they will excel in them. These people will apply intense focus to a difficult task, prolonged and repeated efforts over and over to get to their end goal. This is often referred to as N-ach.

Need for cognition

This is where individuals have the tendency to want to engage in activities that require some mental challenge (puzzle, maths). These individuals have high motivation in cognitive tasks. Just to note one thing people with a high need for cognition pay more attention to arguments in ads.

Regulatory Focus

Individuals with this trait are more promotion oriented, seeking out opportunities for monitory gain. This is more of a goal-oriented pursuit type of trait. We all display some levels of the regulatory focus traits, at some point in our lives, if not I am sure we can identify one of our co-workers who display that type of trait.

Sensation Seeking

This trait is self-explanatory, these persons are motivated to engage in extremely risky behaviours. They are more likely to walk off a job with no plan, engage in unsafe sex, and use a heavy substance for the sole purpose of getting a rush. Sensation seeking trait is few, they are often more extroverted as well, and very open-minded.

Chapter 6
Errors in understanding motives

Based on the information we have from the previous chapter, you may believe that you know can predict any one's personality or behaviour easily, not totally true. As we stated earlier in the book, predicting person behaviour is possible, and practically, however, you may not be 100 percent correct all the time. There are some issues when it comes on to predicting a person's behaviour, which we will look at this at this moment. As human beings often we may be subject to interpretive biases that may obstruct rational thinking, which may misguide us in understanding a person's behaviour. Most of us receive information daily from vast amounts of sources and personal experiences.

From these experiences and source, we develop in our databased an interpretation of the motive based on our past encounters, that holds our judgment of the event. These past events and experience will develop in us an intuitive interpretation which may be wrong sometimes. As each case may be different, it's sometimes necessary for us to make our assessment of a situation case by case.

***Why our assessment of a person may often be incorrect is:* -**

People often don't understand their own motives, our interpretation of motives can be distorted in many instances, as unlike physical attributes, psychological markers such as preferences, beliefs, and disposition cannot be understood at face value. As we will find that most people do not have the same persona as they do in private, we may find that a person may be introverted around new people and extroverted around their friends. Or they may be shy and timid at the office, however, they are studs and go-getter around their school buddies. One of the reasons for this is that there is a phenomenon of social desirability, where a person may display certain behaviours because it's more culturally accepted. So a person will deliberately suppress who they are to meet the social expectation of others.

Behaviours are examine based on personal experiences, thinking that our way is correct and just. The best example for this is driving, when you are driving you find that you are always right. This comparison often leads to flawed conclusions, when we compare others behaviour to ourselves. What we often do at times which further complicate the interpretation process is to focus on the evidence only to support our own desires and interpretation. At the same time, we ignore away significant disconfirming evidence. Interpreting through personal lens only fail to recognize the role that culture, social development and emotional state of the individual you encounter. So often times, we have to make our assessment objectively and consider many factors before we arrive at a conclusion.

Not all behaviour represent the same motives, simply observing a behaviour may not be all there is to arrive at a motive for the behaviour. Because similar behaviour may indicate different motives, the same motives may result in a different response, not everyone responds the same to similar triggers.

People may have multiple and simultaneous motives with some objectives prioritizing more than others, resulting in a hierarchy of motives.

Our ultimate goals are often suppressed in favour of smaller goals. For example, a person may enrol in an adult education course to show their capabilities or they could be there for social reasons. There is always some levels of depth to us analysing a person, as human beings are complex. So for us to make a judgment solely on observation alone, we have to consider multiple strands of contributing factors.

Motives are often combined with personality and character, even though personality and character are more enduring that situational motives. Many times people will conclude that personality trait will always determine motive and intent. Even though personality may be used to predict what a person may do under a specific circumstance. The entire nature of the circumstances may result in a different motive, thus derail the predictive ability of the personality.

You will also have times when other factors will override predicting behaviour pattern, such as when an introvert will notice someone familiar of the same interest in social settings and then suddenly becomes the life of the event.

The process of labelling a person according to motivational type can be risky, considering that motives can be fluid, malleable, and changeable and always fluctuating. Based on different factors such as social environment and additional personality characteristics that may be dominant.

Emotions can disguise or disrupt normative behaviour, emotions can quickly lead to a false interpretation of motive.

When a person is under a lot of stress and emotional strain, their responses often will not reflect their true personality. As the mind often succumbs to the perception of pressure, stress and other psychological and physiological strains by the prevailing state at that time. Their actions may sometimes be outside of our control.

Accurate interpretation of behaviour and personality will require the understanding that all environmental events will be subjectively interpreted across the wide range of individual emotions.

Chapter 7

Secrets of Reading Body Language

It's very interesting that body language says so much about a person personality, it gives us an amazing amount of information. Research has shown that 55% of our communication comes from body language, while only 7% come from words. So you see why it's important to pay much more attention to body languages. We spend all of our lives, learning unconsciously learning how to understand people, learning how to decode them. Everybody you come in contact with is trying to decode you subconsciously. Here are some body language cues that you need to look out for:

Crossed arm and legs

When you see someone crossed their arms and legs, it might be a bit obvious that they are not open to whatever it is that is being said. The crossing of arms and legs reflects barriers that the person has against the suggestions. Psychologically this action signals that mentally, emotionally as well as physically this person is blocked off from who so ever is in front of them. Even if you see them smiling, and looking pleasant, unconsciously they are resistant and you should be able to pick up on that. However the action is not intentional, it's usually deep rooted and psychological.

Smiles and the eyes

We have all given that fake smile at some point in our lives, when it comes to smiling, the mouth can lie, however the eyes will not.

Genuine smiles will reflect in the eye, with crinkle skin, this will reflect in your physiology. People often smile to hide how they truly feel, so you can analyse whether someone is genuine with their smile, by looking at their eyes.

Mirroring Body Language

You may notice at times that you may be in an interaction with a person and they start to mirror what you are doing. Such as when you cross your legs they do the same, when you move your arm, they do the same. This is a sign that the person is unconsciously receptive to your message and your conversation is going well. Having knowledge of this is great, especially if you are negotiating a sale or courting a partner.

Noticing Postures

Your posture tells a whole story about your state, it does not entirely tell us about your personality, because it may be just an experience you had that changed your state. However, if the posture a person has shown is prolonged then we can make predictive judgments about their personality type. An erected postures we can tell that the person is confident and sure about themselves, while a more slump, slouching type of posture would mean that the person may be a bit insecure about themselves. Standing up straight is a power position, this form of posture commands respect and leadership. A collapsed posture commands little to no respect and projects less power.

The eyes of deception

The eyes do say a lot to tell you if a person is being truthful or not, you will often hear a person saying "They couldn't even look me in the eye". That's because uncertainty is first discovered in the eyes, then you will see the emotions in the body language. But not all the time, you will notice by their body language, especially if you're just having a brief conversation with them. On the other hand, a person that is lying sometimes because they are trying to convince you otherwise, they will overcompensate and try to hold their stare longer, so as to dismiss any signs which would indicate that they are lying. Look out for any unusual eye movement if you are trying to discover if one is lying.

Raised Eyebrows

When you see a person raised their eyebrows, it usually is a signal of discomfort, surprised, worry or fear. If you are talking to someone very casual, and they raised their eyebrows it could mean something. What you have to do is be a little more analytical to see what's really causing this reaction. Every single reaction in the body has a cause, it's for a reason, no matter how insignificant it may be, it is for a reason.

Observe the head movement

How fast a person nods their head, indicate their level of patience, slow nodding would indicate that the person is interested in what you are saying. Nodding very quickly indicate that the person is anxious to either end the conversation or have their turn to speak. Tilting of the head sideways may indicate that the person is interested in what you are saying. If you, however, see the head tilt backward, this can be a sign that the person is uncertain of what you are saying or a bit suspicious.

Hand Signals

Hands play a big important in sending nonverbal cues into what the person may be thinking. Hands in the pocket could be that the person is a bit nervous, or anxious about something. If you are in a room having a meeting and you see where a person is unconsciously pointing towards someone all the type while they are speaking, they may share some affinity with that person.

When you are at a table and you happen to see someone supporting their head with their hands, with elbows on the table. This could indicate that the person is focused on what is being said, however, based on the facial expression you can tell that a person may be experiencing boredom.

Hands on Hips: When a person stands with his or her hands on their hips, it's a signal that they are ready and in control, it may also be that they are aggressive.

Hands Clasp behind the back: When you see this happen the person may be impatient, bored or a bit frustrated. You would have to put the situation in its correct context to know which it is.

Personal Space

I am sure you have heard or said yourself that you need some personal space, at times we ourselves feel uncomfortable if a person is standing too close. Just as how body language and facial expression communicate what the person may be thinking, so does the personal distance. The distance between which two-person stands.

There is *Intimate Distance* which is 6 to 8 inches, this physical distance suggest person are in a closer relationship or are very comfortable between each other.

Personal Distance which is 1.5 to 4 feet, this level of distance is usually between family members or really close friends. The closer a person stands between each other, usually reveals their level of intimacy.

What we experience at socializing with people is *Social Distance* 4 to 12 feet, this intimacy level has to do with persons who are a bit more acquainted with each other. Someone who you know fairly well, it may be a co-workers an old classmate or a childhood friend that you are interacting with.

For persons that you just happened to see at the mall this physical distance is normally wider 12-25 feet *Public Distance*. This is normally for public situations, talking to a group of people or a vendor you are making a transaction with.

Chapter 8

Quick signs that someone is lying

Red flags to look for to identify when a person is lying:

- Giving vague responses, providing little details, leave out important parts of the story.
- Repeating questions, before they provide a response
- Not providing specific detail when their story is challenged
- Grooming behaviour such a playing with their hair
- Ask them to tell their story in reverse to get the correct details
- Speaking in sentence fragments, vocal uncertainty.
- Overthinking, you find that they are thinking too hard to give you the details.

Profiling Dangerous People

Dangerous people are often considered to be narcissist, they often carry similar personality trait. Not to be misunderstood, not all dangerous people are the narcissist, and not all narcissist are dangerous people. But often we find that they overlap each other, and thus we are able to identify and interpret if a person may be a potential threat if they display certain type of behaviour.

1. If you find that they expect to be treated special, different from everyone else, or they expect to be the priority at all times, that's a red flag. We all have experienced person like these, especially when we were younger growing up. We often had experiences with the spoil brats or the single child that just feels entitled to the

world. Many of these individuals that feel entitled and special can turn into the worst human beings if they do not get what they want.

2. If you notice that they overvalue themselves, thinking that they are the centre of the world and everything revolves around them. These type of persons are ego driven, they do not want anyone to oppose them or challenge their ideas, they believe it's the end all and be all. They tend to devalue other person contribution, and ideas, they would rather take the credit for everything that was done. Be warned! Many of us have encountered these people at our office, some are our co-workers.

3. If They tend to not care about persons unless they contribute to their intentions. If

you are not there to help them, they have little regard for you, they almost brush you off as if you do not exist. Once they are able to get something out of you, they will then become your partner in crime.

4. They are very poor listeners unless they're at an advantage to gain something out of the conversation. They tend to operate with a sense of dictatorship, whether it is that they are in a position of power or not. Once they are given the room to, they will automatically behave as if, everyone is their subordinates or they are royals and everyone else are peasants.

5. If you are around a person and they always demand that you remain faithful and loyal to them as if be loyal or else? Those are some red flags for you to be aware of. Some person may be a bit insecure about themselves, however by instincts alone, we can pick up on which persons are insecure and which persons are outright dangerous. Once I identify that a person may be a little bit of control freak, I

normally try my best to remove myself from inside that type of relationship.

6. They are never in the wrong, they are always correct it doesn't matter what the circumstances may be. They could have straight up punched you in the stomach they will find some way, to suggest that you deserved it and you are the one to blame. They will do this, without feeling any remorse or regret, no empathy or guilt, if you notice that someone in your immediate day to day contact display these traits, they may be a bit dangerous.

7. If they are in a compromising position they will play the victim role, or the poor thing so they can get special attention. At times the situation may not even be compromising and they will always play the victim role to get

attention. For them to again can be the centre of attention at all cost.

8. They make a mountain out of a pickle of the situation, meaning they become overly angry at the minute circumstances. An issue that can be easily blown off, they get overly aggressive and irate at stages where it may even be unbelievable. They turn 5 minutes arguments into hours, prolonging something that is very insignificant.

9. They are always suspicious of people, they believe that someone is out to take advantage of them or exploit them. Even good intents are misunderstood for exploitation, they have a paranoid personality.

10. They are always on guard, always up to something and believe that others are of the same way. They project their destructive

thinking towards others, they believe everyone is the same. They themselves can't trust anyone, and they should not be trusted just the same.

Chapter 9

Judging a book by the cover

We can tell a lot about person from the look on their faces, our minds automatically can read and interpret a persons expected personality. Just think about it, how many times have you seen a person and think, this person must be kind and gentle person, which they are. Then in other instances you will see another person and think that they look aggressive and unapproachable, which they often turn out to be that way.

As we discussed earlier this has to do with the baselines that you set earlier, scientist call this practice physiognomy. Reading peoples character from their faces, faces do deliver important information about someone's personality. We have all heard the expression that our eyes are a window to our soul, this is quite true in fact. Our faces and eyes really does say a lot about us.

Wrinkles in the face tell a story, it goes beyond just telling us that we are getting old. We use different facial muscles when we smile, when we are depress, angry and sad. What happen is often, when we are in these states our facial muscles for these states are shown more even when we are calm and thinking about nothing.

If you have crow's feet at the corner of your eyes or even your lips, this indicates optimism. It shows that you love to laugh. Just the same if your forehead creases it says that you are a serious person and have endured a lot during your lifetime.

Let's get a little nosey, scientist have discovered that if, you group several persons together, the bigger the nose is often the more audacious the person is. These people are often very competitive as well have higher standard for performance. Scientist have also discovered that in terms of investing, if the investor nose is pointing and edgy, they are often excellent investors, if the tip is small and delicate the owner is more reserved with their finances.

These insights can be misleading at time, however it serves as a guide for us to make a general assessment of an individual's personality. For example a person with angelic face can also be a child molester or even an alcoholic. Alexander Todorov, had students assess the appearance of person's faces, he discovered that the general view was that people with round faces, with big eyes appeared to be more fragile, obedient and sincere. While a ugly more gloomy face was viewed as having an unfriendly personality.

Chapter 10

Emotional Bullying, How to spot it!

Bullying is something that some of us have experienced in our childhood, it may have been verbally abusive or physically abusive. We internally feel the effect of both, emotional abuse as an adult can take on another form. We can experience this from our spouse, friends, co-workers even family members. Emotional abuse affects the individual negatively, at times persons themselves may even turn around and become bully's themselves.

Victims can feel guilty, shameful, disrespected and embarrassed. The effects of emotional bullying can take the form of depression, shyness, lack of confidence, low self-esteem etc.

If you feel insulted by a person, feel as if you have to talk to the person careful or be quite around the person that means you may be suffering from emotional abuse. You can recognize emotional abuse when the relationship constantly makes you feel bad about yourself or lowers your self-esteem you are in an emotionally abusive relationship. This does not only occur in an intimate relationship, but it also occurs in relationships among friends, family members, and even co-workers.

Emotional abuse can very subtle, almost unrecognizable to a person, sometimes this will result in one finding it hard to identify emotional abuse. If you are unable to identify whether you are encountering emotional abuse or not, take a stock of how you feel after an encounter with them. The abusive person always makes unrealistic demands, expecting you to spend all your time with them, and no one else. It doesn't matter what other obligations you may have, they want you to drop everything and focus on them. They feel as if, they are entitled to your time, where you should give them it all.

How to spot insecurity

So many times we are interacting with a person and we sense that they may be a bit insecure, and at times we don't even recognize insecurities. But once we understand how to recognize and identify insecurity, we are able to then sense what could be contributing to them being insecure. For example, you may be in a relationship, and you decided to go out, if you can identify that the other party is insecure, you will then know that, your spouse doesn't trust you. Another example if, you are at work, and a new employee is introduced and you notice how maybe some of your co-workers are insecure, you can then tell that they may feel threatened by their position.

So just the simple process of identifying when a person is insecure, you can then through deductive reasoning understand why. So let's look into some ways to spot out insecurity.

1. If you are around a person and it makes you question your own self-worth, it could be that the person you are dealing with is projecting their own insecurities towards you. An insecure person will try to make you insecure yourself, they will make you feel insecure at your job, insecure about your body, insecure about your spouse. At times, it may be toxic, and you may need to dissociate yourself from these persons. They will say things like, I hate this job, or I think my spouse is cheating, or I think I may be fat and unattractive. When you are surrounded by these type of input you

yourself become insecure about the same things.

2. If they are always talking about what they have accomplished, what they have achieved, who they know and what they have done, that person may be insecure. An insecure person always wants to put their guard up, by trying to convince other persons that they are themselves are worthy or valuable. If you don't believe this, the next time you see someone talking about themselves and their accomplishments, just walk away. It may be rude, but try it, and you will see that they will be crushed, why? They value how highly you think of them, they want you to see them a certain type of way as if they are on a pedestal. That is a direct demonstration of insecurity.

3. You will always hear an insecure person say that their not good enough, even if they are great at what they do. Often people that like to display high inferiority openly like to show off how well they can do a something. I am sure you have experienced this at your job, in school or maybe you yourself. They may often time come over as snobs, the fact that they are putting on an act that does not convince anyone, them knowing that they set higher standard for themselves.

Chapter 11

Analyzing People by Birth Order Behaviour

First Born

The first born is normally treated with more attention and care solely because of the fact that parents are usually excited about their first born and do not want to make any mistakes. With this you will notice often that the child will seek to be perfect, to appease their parents. When parents raise their first child, they often raise them with an iron fist, strict and disciplined, this has a significant impact on the personality and behavior of the child.

Firstborn is often the leader of the pack, they are reliable, conscientious, controlling and are high achievers. They want to be the best dressed, best employee, they love to be liked and are often extroverts.

When you are in a conversation with them, you will often hear them say, I want to have the best company, I want to create the best team, I want to be the sexiest, I want to be the fastest, fittest. These are some of the cue statements you can look out for to identify if someone is the firstborn person of their family. Your analysis may not be correct 100% of the time, however, the probability is really high that you will get it correct most of the time.

Middle Child

The middle child is normally the one that feels left out, given that most of the attention from parents are on the baby in the family or the firstborn. The middle child will often seek to find their place among their peers, they will often turn out to be people pleasers. The middle child are often rebellious, embrace friendships, they have a large social circle and are often the mediators in any disputes. People usually love to have them in their company because of their personality.

The middle child will often be the ones that are pushing the envelope, taking a risk and going above and beyond. They are generally rebellious and love to have their own way despite the odds or the consequences. They are extremely social beings and are a very good team player.

The middle child can be very resentful and bitter if you catch them on the wrong side, however, they do display strong leadership qualities. Despite the fact that at times they may be stubborn and hard to work with, they are excellent to have around as friends.

Last Child

The youngest child would be more liberated and free-spirited because of the fact that parents are less strict and stern when they are dealing with these children. These people tend to be more adventurous, thrill seekers, outgoing, extraverts and attention seekers. The last born child is usually careless and unorganized, they are impatient risk takers. These are the ones that are generally able to get up and walk off their jobs without having a backup plan. This makes them very competitive and ambitious, due to the fact that as the youngest sibling they often see the older ones doing things which they can't.

They often do whatever it takes to get their goals, get them as urgent as possible. In a relationship they expect a lot of attention and love, if they have a spouse that does not understand that they need attention, this will turn into a problem. Because of their birth order, they are used to receiving lots of attention and love early in their life and expect the same.

The Only Child

The only child is considered as the "little emperor", they are high achievers and more often successful. They would not have to compete for attention, they receive all the love and care. They may be show-offs and behave a bit cocky to maintain the attention he seeks. They have great social skills and networking skills because they had to spend most of their time around adults.

Confidence is not an issue, public speaking is their strong point and they know how to manage and enjoy their own time. Their personality is a bit complex, remember I said that these traits are not set in stone, because there are many factors which make up a personality. However, there are certain traits and expected behaviours which are consistent and predictable.

Chapter 12

How to detect romantic attraction

If you are in a social gathering, a party, a club or even at your workplace and you spot someone constantly gazing in your direction. Chances are that person may be sending you signals that they are interested in you. In particular, if a woman is attracted to a man, most of the time, she will gaze at him several times until he notices her. It's up to that male to pick up on the signals being sent and make his move. For men it's almost the same, however, what men will do is try and hold to gaze on the female until she notices him, then afterword's he will try and make is move if there is interest.

If women are interested in a male, they will often tilt their head a bit in the man's direction or you will see the fixing their hair a bit, or orient their body to face the other person. Further sexual attraction will result in more soft touching between each other, you will see a synchronization in gestures more head tilting and leaning forward into each other. Women have to pay more attention to how a man will react to them, especially when they are talking about their personal lives. If the guy you are talking to enjoys listening to you talking about your personal history or just discussing science, politics or world affairs subjects, this can indicate he is attracted to you.

Secrets of the handshake

How many times have we shook someone's hand and instantly we can tell some aspects of their personality. A psychologist has found that how you shake hands, alters people's impression of you.

People with firm handshakes and strong eye contact, tend to be better performers and more confident individuals. They are more consistent with what they do, they are extroverted and positive more than others. While someone with a loose handgrip is more often shy, introverted and experience a lot of social anxiety.

Putting it all together

The human mind and behaviour is quite complex and challenging to understand, the mind is still more powerful that any super computer on earth, think about that for a while. Then think about how challenging it is, for you to understand a supercomputer, that is the challenge we encounter with understanding people and their behaviour. But we can learn how to analyse and pick up on cues to help us in better understanding, behaviour trait and personality.

All this is done through correct information, information we use to guide us in our understanding and interpretation. The information that I have provided inside this book will make you less susceptible to falling a victim in an encounter with someone.

Instantly once you begin to practice reading and understanding people, it will then become more automatic without you consciously thinking about it. You will find yourself picking up triggers and cues, and looking out for follow up behaviours, words and actions. Be reminded that not everyone is the same, not all interpretation will be accurate, but we can get as close as possible to analysing and interpreting personalities correctly.

I know this book will help you significantly as well, this will allow you to also teach some of the lessons to your friends and family. I hope you had fun with this book.

Once you are through with reading or listening this book, please remember to leave a review it would be much appreciated. Thank you!

www.ingramcontent.com/pod-product-compliance
Lightning Source LLC
Chambersburg PA
CBHW030704220526
45463CB00005B/1895